Safari in the Bay of Biscay

~~~~~~~~

Andy Williams & Tom Brereton

TSG Publishing
P O Box 8
Romford
Essex
RM4 1UY

First Published in Great Britain 1998, by TSG Publishing
First Reprinted in Great Britain 1999, by TSG Publishing

Copyright © Andy Williams & Tom Brereton 1998
Copyright © Photographs Andy Williams, Rolf Williams,
A&R Williams & BDRP 1998
Copyright © Biscay Dolphin Research Programme
All rights reserved.

Without limiting the rights under copyright reserved above, no part of this publication may be reproduced, stored in or introduced into a retrieval system, or transmitted, in any form or by any means (electronic, mechanical, photocopying, recording or otherwise) without prior written permission of both the copyright owner and the above publisher of this book

Printed in England by LJ Print & Design, Yeovil, Somerset.

CHAPTERS

~~~~~

Introduction

Chapter 1	Description of the Area
Chapter 2	Whales, Dolphins and Porpoises
Chapter 3	Description of Behaviour
Chapter 4	Dolphins of the Channel and Bay of Biscay
Chapter 5	Whales of the Channel and Bay of Biscay
Chapter 6	Birds of the sea crossing
Chapter 7	Wildlife around Santurtzi
Chapter 8	Marine Conservation and the Biscay Dolphin Research Programme

INTRODUCTION

By Andy Williams

The area of sea that is bordered by the north coast of Spain and by the southern coast of the Brest Peninsula has achieved an almost legendary status amongst seafarers. This area known as the Bay of Biscay has for many years been thought of as a dangerous and inhospitable part of the world's ocean system. To many people fishermen are heroes for braving the mountainous and unpredictable seas, indeed there is still a certain romance associated with a seafarers lot.

Ask many of those same seafarers and they will tell you that 'the Bay' is an area where there are few whales and dolphins and not much else other than tuna.

In 1995, I approached P&O European ferries to see if they would be interested in allowing me to run a research project based on the ship. The idea was to travel monthly and to place observers on the 'Pride of Bilbao' operating from the bridge. The ship had been plying its trade from Portsmouth to Bilbao for the previous couple of years. I received a very favourable response from P&O European Ferries (Portsmouth) and the first trip took place during August 1995.

The first trip also included a BBC film crew from the children's news and magazine programme, 'Newsround'. One young girl, Zoe Barlow-Hill had written a story about the Bay of Biscay and her prize was to travel with me on the very first trip. I had expected to have at least one other person with me, someone who was an expert in this type of survey work. Unfortunately at the very last minute he was unable to come and I was left to do my best with the survey, the TV crew, and Zoe who was a very interested young girl.

At the end of the trip we were doing some filming and I said how lucky we had been to see the animals we had. During that trip we had seen 3 different species of whale and dolphin and had recorded 15 sightings. I had no idea at that time what I would see or what I would experience over the next two and a half years. I gathered together a small core team, Tom Brereton and Rolf Williams and between us we have carried out the survey every month since August 1995 using occasional observers in addition to the core research team to assist.

I was increasingly surprised by the trips we made. I read a great deal of the available literature about the area and gained a reasonable picture of the types of animals we might see. However, nothing had prepared me for the reality that was slowly unfolding before our eyes on each trip.

To our eternal regret we did not start carrying a camera capable of taking the type of photographs we needed until we had missed a great number of exciting and spectacular sights. This was because we were (and still are) all volunteers and we received no funding from anyone for the work we were doing so expensive cameras and lenses were a luxury we could not afford.

Things changed as the crew, particularly the entertainment team took our work, and us, to their hearts. They began collecting money through a raffle held amongst the passengers on each trip and making donations to the project. Having money available allowed us to create the Biscay Dolphin Research Programme (BDRP). The money the crew raised and continue to raise keeps the project alive and their contribution is invaluable.

The enthusiasm of the passengers amazed and excited me and I was keen to show everyone the sort of whales and dolphins they could see on the crossing. I firmly believe that science is something that everyone should be excited by. It belongs to the people and it is the job of those who carry out research to ensure that the mass of the population are informed and excited by those things we find out.

To that end and with some of the money raised by the entertainment crew I had the idea of a display board on the top deck telling the passengers how to watch for whales and dolphins and what they might see if they put in the effort. Rolf managed to recruit Dorling Kindersley to help with the design we had drafted.

The board has proved to be a great attraction in its own right and even on wet and cold days we still see people reading it. I hoped that the display would help people to interpret what they saw and I have seen it work many times since it was unveiled.

The project has been undertaken to try and find out what animals use the Bay of Biscay and how that changes during the seasons. We expect to travel on the ' Pride of Bilbao' for a period of 5 years to get a good picture of the area. We are not only recording the whales, dolphins and porpoises in the area but also birds, sharks and sunfish.

This book will, I hope give you an insight into the animals and birds that you might encounter if you travel across the Bay of Biscay. I consider myself to be one of the luckiest people on earth because I am able to travel each month and watch this fascinating area change character and to look upon animals that most other people will only dream of seeing. I hope through these pages you may start to view the bay as an exciting and vibrant place to be. Should you be lucky enough to see some of the life in the bay I hope this book will help you to identify them and that you will learn something about what you have seen.

MAP OF AREA

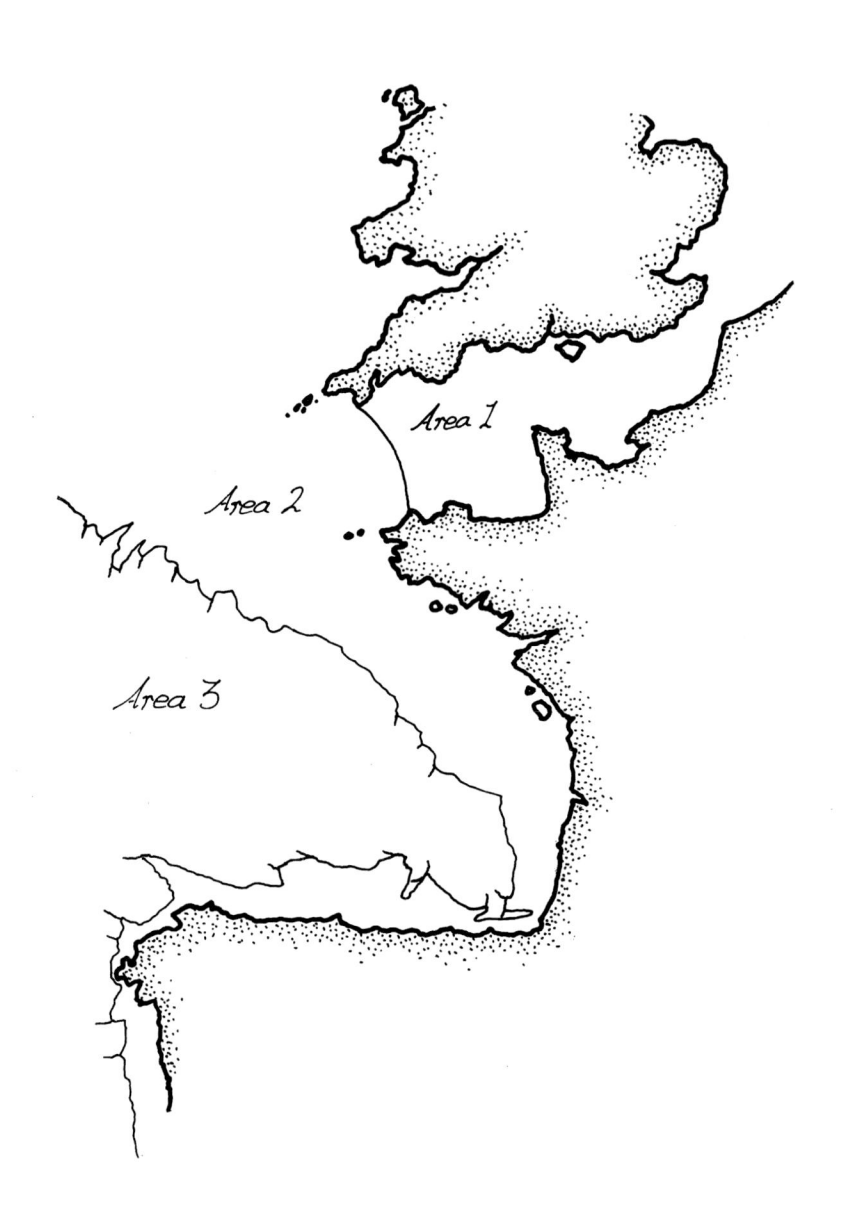

Chapter 1

Description of the Area

The area that is described in this book can be divided into 3 parts. The areas each have a different profile and the species that are found are often associated with one or two of the different habitats.

Area 1:
Area 1 is the Channel and is bounded for our purposes by the Lizard Peninsula to the north and the French coast at the Brest Pennisula in the south. This area of shallow seas with the depth typically less than 80 metres, but with deeper trenches reaching depths of up to 110 metres.

Area 2:
Area 2 is the area of continental shelf that forms an edge around the deepwater area of the bay. The depth in this area varies between 110 and 800 metres. The continental slope provides an area rich in nutrients and therefore rich in the prey of the dolphin species in particular. The angle of the slope varies wherever one crosses the shelf. In the north during the 'Pride of Bilbao' surveys we cross the shelf at a point that has a relatively gentle slope. On the return journey the shelf plunges away to the deep Biscay abyssal plain.

Area 3:
Area 3 is the area from the upper continental shelf to the deep Biscay abyssal plain. The depth in this area goes from 800 metres to 4529 metres. This is the domain of the deep diving whales and the oceanic dolphins. It is thought that the edge of the shelf is also used by some of the migrating whales as a route guide. The

abyssal plain is home to the giant squid favoured by sperm whales and the deep-water fish and squid sometimes eaten by the beaked whales.

Each species of whale and dolphin live in a particular habitat and are adapted for that environment. Many species have a diet that allows them to eat whatever is around: others may need to be more specific. Each of the whales and dolphins have evolved to fill a niche in the marine environment, they may cross the areas we have described but the areas provide a guide to the variation in sightings as the Bay of Biscay is traversed.

Chapter 2

Whales, Dolphins and Porpoises

Whales dolphins and porpoises are collectively called cetaceans. During the first three years of the survey carried out by the Biscay Dolphin Research Programme, 20 different cetacean species have been observed and recorded.

Whales can be classified into two groups; the toothed whales and the baleen whales. Toothed whales include, all the dolphins (including the killer whale), and other members include beaked whales, pilot whales and sperm whales.

Baleen is a hair-like substance that some of the world's largest whales use to sieve food from the water. All of the world's largest whales are baleen whales and include the blue whale, fin and sei whales. These whales feed by taking very large amounts of water into their mouths and then forcing the water back out of their mouths making sure that it passes through the baleen to catch the fish or other animals in the mouth of the whale.

We have gathered a great deal of evidence to demonstrate that the species seen in the Bay of Biscay varies with the seasons. Some species are migratory and are seen only as they travel through the bay on their way to, or from summer or wintering grounds. For example sperm whales may make a mammoth trek each year from the Azores to the Arctic and back. Other species, such as the blue whale make even longer journeys and may travel around the entire globe. It may be that yet other species move in and out of the bay into the deeper mid-Atlantic waters.

Some of the smaller dolphins like the striped dolphin, move into the south of the bay as the water temperature cools and winter approaches. As the spring arrives and the water temperature increases so we start to see those species again. By now these animals have young with them and the groups we have observed are often larger when there are new-born animals within the group.

Larger groups can often be used as protection for the young, weak and vulnerable. For whales and dolphins dangers exist from other cetaceans such as killer whales, melon-headed whales and pygmy killer whales. Sharks also represent a threat to those least able to care for themselves.

As spring arrives so the calves appear in the area, fin whale mothers move into the bay with their calves. The calves are nearly the size of fully grown killer whales when they are born. Calves are also seen in the late summer in some numbers and again large groups aggregate.

Cetaceans may travel many miles in a day and many thousands of miles in a year. We are lucky if they choose to spend some of that time within our view. Although whales and dolphins are air breathing mammals they spend a small fraction of their time at the water surface. When they are at the surface they can display a wide variety of behaviours, some of which are characteristic of that species or of just a few species.

Chapter 3

Description of Behaviour

Cetaceans are complex animals that display a large variety of behaviours for the observer to ponder. Not all species display the same behaviour and some are spectacular and acrobatic, others subtle and sensitive. This section describes some of the more frequently observed behaviour that one might see and the species, which display this behaviour.

Bow riding: common dolphin, striped dolphin, bottlenose dolphin, white-beaked dolphin and white-sided dolphin.

Bow riding animals often approach the ship or boat at speed and then remain in an area just ahead of the vessel moving without effort.

When observed they appear to hardly move their tails and yet they always stay ahead of the vessel. This is because they are sitting in an area of water (called the pressure wave) that is constantly being pushed ahead of the ship.

Some larger ships have a bulb on the bow that act as a very good pressure wave production mechanism.

Aerial displays: striped dolphin, common dolphin, bottlenose dolphin, white-sided dolphin, white-beaked dolphin, and Risso's dolphin, Cuvier's beaked whale.

Aerial displays are a way of describing those things that whales and dolphins do naturally, that are artificially induced in dolphinaria using props such as balls or people with fish in their mouths. Both whales and dolphins leap high and clear of the water. Some will somersault and others will leap vertically out and then slap down back onto the water on their side. Some species of whale including the Cuvier's beaked whale are known to do this. It is quite a wondrous sight to see a whale of 6 metres or more in length clear the water by some 5-7 metres.

Breaching: Minke whale, humpback whale, fin whale, killer whale, sperm whale, bottlenose dolphin, Cuvier's beaked whale, juvenile pilot whale.

In whales, breaching occurs when the animal rises out of the water in one fluid movement, does not clear the water completely and then splashes back into the water. Breaching is a quite astonishing sight particularly when a 20 metre animal is half clear of the water before crashing down on the water surface creating an explosion of spray.

Lob-tailing: Risso's dolphin, pilot whale, sperm whale, humpback whale and killer whale.

Lobtailing is a strange sight when first encountered. Lob-tailing animals will lay in the water vertically with their tail flukes and in some cases most of their tailstock clear of the water. The tail can stay out of the water for some time.

Spy-hopping: Risso's dolphin, pilot whale, killer whale, humpback whale and sperm whale.

Spy-hopping but like lob-tailing involves the animal lying vertically in the water, this time with the head and part of the body is out of the water. It appears that they are watching us watching them.

There are many other behavioural patterns, that one might observe in animals whilst crossing the bay. Whale and dolphin behaviour is a very complex and intricate part of their communication system. We are starting to get an insight into some of those behavioural patterns but they are only one part of a huge jigsaw.

Chapter 4

Dolphins of the Channel and Bay of Biscay.

Harbour Porpoise:

Distribution: Areas 1, 2 and 3.   Abundance: Common.

Group sizes vary but are generally 2-5 with larger gatherings at feeding areas. Much less common in the offshore areas and the areas around the Iberian Peninsula.

Birth size: 0.67-0.90m(2.2-3.0ft)   Adult size: 1.4-1.9m(4.75-6.25ft)

The harbour porpoise is the smallest cetacean in the north Atlantic. It is recorded regularly in the Channel area but is difficult to see from a height of 30 metres because of its small size and lack of acrobatics. The porpoise can be identified because of its small size and its small triangular dorsal fin.

This animal has suffered in many parts of the world because of the bottom set gillnet fishery. Porpoises are bottom feeders and they seem to be unable to comprehend a very thin fishing net that is attached to the seabed and they become entangled and drown. There is a great deal of research effort being put into finding a solution to this problem. Porpoises are generally found in the shallower areas and have been recorded in the approaches to Portsmouth Harbour.

Common Dolphin:

Distribution: Areas 1, 2 and 3.   Abundance: Most commonly seen of the small dolphins.

Group sizes: Between 5 and 1500.

Birth size: 0.80-90cm(2.6-2.9ft)   Adult size: 1.6-2.6m(5.3-8.6ft)

The common dolphin is the species seen most often in the Bay of Biscay during the ship's passage. It has been recorded on a number of occasions in the Channel and western approaches. This species is very energetic and often comes to ride the bow wave of the ship. The group sizes have been seen to vary with the seasons from an average of 10 animals per group in February to 170 in June. They are the most colourful of the dolphins with the bold yellow band on the flank distinguishing them at close quarters from any other species.

This is the species that have been recorded most often during the first three years of the BDRP. Common dolphins have been seen on more than 200 occasions with the largest single group comprising 1,500 animals.

Striped Dolphin:

Distribution: Mainly Area 2 but also recorded in Areas 1 and 3.

Abundance: More common in the deeper waters and in the warmer southern waters but can be seen further north during the summer.

Group size: Recorded in groups up to 100. The striped dolphin has been seen in small groups with common dolphins.

Birth size: 1.0m(3.3ft)   Adult size: 1.95-2.4m(6.4-7.9ft)

Striped dolphins can often be seen at distance because they leap high into the air carrying out acrobatic feats that may see them leap more than 5 metres. When seen in large groups they often come to within a few hundred metres of the ship but bow ride or play in the stern wash much less than common dolphins.

Striped dolphins have a seasonal movement in the bay area. They are seen during the summer months throughout the cruise but as winter approaches, they are seen less frequently in the north of the bay. During the winter striped dolphins are only occasionally seen in the far south of the bay. Striped dolphins are amongst the most frequently seen cetaceans in the survey area. They are always a pleasure to see and their antics will hold observers fascinated and bring a smile to their faces.

Risso's Dolphin:

Distribution: Areas 1, 2 and 3.   Abundance: The population of Risso's dolphins is unknown, it occurs around headlands and islands and in deep waters. There may be populations that keep to an area but little is known about these animals.

Group size: Often seen in groups of between 3-50 but there may be 150 or more in a social group spread out throughout the range.

Birth size: 1.2-1.5m(4.0-4.9ft)   Adult size: 2.8-3.3m(9.2-10.9ft)

The Risso's dolphin is a curious species to encounter; their markings vary a great deal. The young animals are grey. As they get older so Risso's dolphin start to become scarred and these scars retain a white colour. Eventually they can become almost completely white. They have no obvious beak.

They have been observed spy hopping and lob-tailing and are often seen tail slapping and breaching. They have very large dorsal fins that are quite erect. The Risso's dolphin seems to associate fairly freely with other species such as pilot whales.

White-Beaked Dolphin:

Distribution: Areas 2 and 3.   Abundance: Generally a species found in the more northerly seas of the north Atlantic. The white-beaked dolphin was recorded in the area around the continental shelf at the north end of the Bay of Biscay.

Group size: 2 - 30 large groups have been recorded of up to 1,500

Birth size: 1.2-1.6m(4-5.3ft)   Adult size: 2.5-2.8m(8.25-9.25ft)

White-beaked dolphins are so called because they have a distinctive white beak that is readily observed from above. They are a very big looking animal and have a large girth. They have been recorded from the 'Pride of Bilbao' travelling in a group with common dolphins. White beaked dolphins are generally found further north than the bay but as with other species they may travel long distances.

Atlantic White-Sided Dolphin:

Distribution: Area 2.   Abundance: Atlantic white-sided dolphins are in the Bay of Biscay or south of the Scottish Isles. However, they have been recorded in the area of the continental shelf in the northern Bay of Biscay.

Group size: 5-50 with schools of up to 1,000 recorded offshore. The first records from the 'Pride of Bilbao' were of a pair of animals.

Birth size: 1-1.3m(3.6-4.25ft)   Adult size: 1.9-2.8m(6.25-9.2ft)

The Atlantic white-sided dolphin is a large dolphin that has a very thick tailstock. It has a yellow or mustard coloured patch on its tailstock. This is a fairly acrobatic animal and when viewed from a distance leaping from the water the white band below the dorsal fin can be quite distinctive.

Their large size makes them readily distinguishable from the common dolphin that has an ochre coloured patch with a white flash below it. This species is also a lot less agile than the smaller common dolphin.

Bottlenose Dolphin:

Distribution: Areas 1, 2 and 3.   Abundance: Fairly common

Group size: In the Bay of Biscay area have varied between 3 and 60

Birth size: 0.98cm - 1.5m(3.2-4.3ft)   Adult size: 2.2m - 4m(7.3-13.2ft)

The bottlenose dolphin is the largest of the dolphins regularly seen from the 'Pride of Bilbao'. It has been recorded in all areas of the survey with both offshore and coastal populations occurring. Studies elsewhere have identified differences between offshore and coastal species in terms of length and girth, but these are not easy to see at sea even for the trained eye. Another feature that distinguishes the two sub-groups is the depth of water they occupy, and so those seen in Area 3 are more likely to be the offshore type.

The bottlenose dolphin is the species most often used in dolphinaria around the world. They have complex social structure and travel widely. Bottlenose dolphins have been observed on a number of occasions in close association with pilot whales and striped dolphins.

Chapter 5

Whales of the Channel and Bay of Biscay

Pilot Whale:

Distribution: Area 1, 2 and 3.   Abundance: A common animal found in all areas.

Group size: Pilot whales are very social animals and are usually in groups of between 10 and 50. Larger groups may be found occasionally and consist of over a thousand animals.

Birth size: 1.5-2m(4.9-6.5ft)   Adult size: 3.8-6m(12.5-19.25ft)

Pilot whales are one of the most social groups amongst the whales and dolphins. They are the species most commonly involved in mass strandings around the world. The causes of mass strandings are still open for debate but there is a large body of evidence to indicate that pilot whales mass strand when groups follow an ill animal that beaches itself. This is likely to be a socially driven response and not a mass suicide as some might try to suggest.

There are a number of rescue organisations around the world that have experience of this type of mass stranding. The most well known is Project Jonah in New Zealand, which has had great success in returning the majority of animals safely to the sea once the seriously ill individual has been identified and dealt with.

Pilot whales are often seen in company with other whales and dolphins. Sometimes they can be difficult to pick out if they are amongst bottlenose dolphins but pilot whales have a blunt head and a very rounded dorsal fin. Care is also needed to separate juvenile pilot whales from bottlenose dolphins, which have a similarly angular and pointed dorsal fin. Pilot whales are generally slow moving and sluggish but juveniles have been seen to leap clear of the water in a surprising display of agility.

Killer Whale (Orca):

Distribution: Have been recorded in all areas 1,2 and 3.

Abundance: A fairly common whale that is distributed widely throughout all of the world's oceans but is infrequently recorded in the Bay of Biscay.

Group size: Generally between 3 and 25 but occasionally groups may number up to 50 animals.

Birth size: 2.1-2.5m(7-8.25ft)   Adult size: 5.5-9.8m(18-32ft)

Killer whales are the largest of the dolphin family; they are spectacular in their habits, beaching themselves in various parts of the world to catch seals. Killer whales have a very complex social structure and there is evidence that they have different languages or dialects amongst groups depending on their social structure and their home range.

The males are very distinct with enormous dorsal fins relative to the female. During the October 1997 survey a group was encountered and a male with half his dorsal fin missing was seen and photographed(see plate).

There are researchers around the world that have catalogues of killer whale dorsal fin pictures and hopefully if these individuals have been photographed before we may get an idea of the extent to which this group has travelled.

Cuvier's Beaked Whale:

Distribution: Area 3.   Abundance: The population status of these whales is unknown but they are thought to be widely distributed throughout the world. They are rarely seen from ships and so they remain something of a mystery.

Group size: In the Bay of Biscay they have been recorded in groups of 1 – 4, although elsewhere in the world groups of up to 25 have been recorded.

Birth size: 2-3m(6.5-9.75ft)   Adult size: 5.1-7m(18-23ft)

Cuvier's beaked whales are curious looking animals. When looking down on fully-grown adults you will see a distinctively brown whale with a large white or cream area extending from the head to about a third of the body length and white or cream spots over the body.

Males have a larger white area than females, whilst the young are almost uniformly grey.

The lifestyle of Cuvier's and other beaked whales is something of a mystery. They are thought to be very deep diving whales that feed on squid and fish from the dark depths of the abyssal plain. Recent reports claim that they may be suffering from the effects of man made sounds in the water. Unfortunately they are difficult to find and more difficult to record.

BDRP has seen as many of these animals as any other research project in recent times and they are truly a fascinating species. We hope to be able to find out more about this shy and elusive whale as time goes on.

Northern Bottlenose Whale:

Distribution: Although it has been recorded in Area 1 these animals are mainly found in the deep ocean areas of Area 3.

Abundance: These whales are not thought to be common but are often seen during the crossing of the Bay of Biscay. The global population size is unknown.

Group size: During the first year of study they were most often seen in groups of 2-5 in the Bay of Biscay, but they have been recorded of up to 35 elsewhere.

Birth size: 3-3.5m(9.25-11.5ft)   Adult size: 7-9.8m(23-32.3ft)

Northern bottlenose whales are inquisitive animals that often approach ships. They may be seen diving under one side and surfacing on the other. It is not known what causes this apparent attraction to ships. These whales are often found in areas where there are deep ocean trenches and they are reported to dive for 1-2 hours. Northern bottlenose whales are usually a brown colour, with a much darker fin than the rest of the visible body.

These whales are also part of the beaked whale family. They were once commercially hunted and suffered large population declines as a result. Recent research in the Bahamas has shown that these animals whilst being very deep diving have a tendency to occupy a definite area of sea. Whether this is true for all

Northern bottlenose whales or just that group studied is unknown.

Northern bottlenose whales, like the sperm whales have an area at the front of the head, called the spermaceti organ. This organ helps them to dive to great depths without harm to their skulls and brains from the intense pressure of the deep ocean surrounding them.

Sperm Whale:

Distribution: Area 3.   Abundance: The world-wide population of the sperm whale is unknown, but there is some evidence that since the whaling ban, numbers have begun to recover.

Group size: Sometimes seen alone but migratory groups may number hundreds.

Birth size: 3.5-4.5m(11.5-14.75ft)   Adult size: 11-18m(36-59ft)

Sperm whales are found on the edge of the continental shelf in the deep waters of the central bay and are occasionally seen off the north coast of Spain. There is an annual migration of sperm whales when groups of predominantly juvenile males known as 'bachelor groups' travel to the polar waters of the north. Sperm whales have the ability to dive to great depths and for great durations; dives may reach 3,000m and last for 2 hours. When a sperm whale dives it often lifts its tail flukes out of the water.

Sperm whale have a hatchet shaped head that is familiar to us from history. It is the animal that was being hunted in Herman Melvilles' book 'Moby Dick' and those pictures of whalers in small rowing boats fighting the mighty leviathan are the classical image that many of us associate with whales and whaling.

Sperm whales can be difficult to see because of their lack of a tall dorsal fin and have tendency to 'log' at the surface. Logging is when a whale lies motionless at the surface for long periods without diving. It is just a resting strategy for deep diving animals. The first indication of sperm whales is most likely to be a bushy, powerful forward pointing blow, which may be surprisingly close to the vessel.

Minke Whale:

Distribution: Areas 1, 2 and 3.   Abundance: A common whale that is normally found at higher latitudes during the summer and lower latitudes during the winter. Found both close to and offshore sometimes in harbours, bays and estuaries.

Group size: Often seen alone or in small groups but may gather in larger numbers in feeding areas.

Birth size : 2.6m(8.6ft)   Adult size: 6.9-8.5m(22.8-28.0ft)

One of the smallest true whales and the smallest of the rorquals (whales with large numbers of throat grooves). Around the UK minke whales are common off the Western Isles of Scotland.

They have a white band around their pectoral fins that is often visible at close range. Minke whales rarely display a blow and if seen, it is weak and very low compared larger whale such as the fin whale.

Humpback Whale:

Distribution: Areas 1, 2 and 3.   Abundance: Rarely seen in the European Atlantic waters but does occur in the coastal waters mainly during April – September. The world population is believed to be around 12-15,000 but the eastern North Atlantic is thought to have only a few hundred animals present.

Group size: 1-3 Larger groups may gather at feeding grounds.

Birth size: 4-5m(13.25-16.5ft)   Adult size: 11.5-15m(37.9-49.5ft)

Humpback whales are included in the group of whales known as rorquals, and are very distinctive when seen at close quarters. They have very long white pectoral fins that are highly visible through the water. When diving these whales will lift their tail flukes clear of the water and each animal has a pattern of markings on the underside that can be used to individually identify them. Humpback whales have been seen in the English Channel (off Portland) and in the Clyde river during the last few years but they are very rare in UK coastal waters.

Humpback whales are one of those species that suffered huge losses during the time of the whaling industry but is starting to recover. In 1989 a leading authority on whales and dolphins reported that as few as 400 were left in the northeast Atlantic Ocean. With that in mind we consider ourselves very lucky when we see one.

These whales produce a wonderful song thought to be the males courting calls to the females. We are still trying to understand so much of the complexities of the life of all the whales and dolphins that these melodious, rhythmical songs provide us with another tantalising taste of the life and times of these travellers.

Sei Whale:

Distribution: Area 3.   Abundance: Officially recognised as a vulnerable species of conservation concern, but no data exists on numbers in the eastern north Atlantic area.

Group size: This whale is normally recorded in groups of between 1-8, but groups of up to 30 may gather at feeding grounds.

Birth size: 4.4m(14.5ft)   Adult size: 13.6-14.5m(44.9-47.8ft)

Another of the great whales of the rorqual family. This is again a very large whale but is significantly smaller in an adult size than the fin whale. The sei whale is another species that suffered seriously through whaling. Population estimates are unreliable for this species. Like the fin whale its blow is often the first sight of the whale that is seen. Given a calm day with little wind the noticeable thing about a sei whale blow is that it is lower (3m) and bushier than a fin whale.

The dorsal fin of a sei whale appears noticeably larger and more sickle shaped than that of a fin whale. Sei whales are seen much less often than fin whales, this could be for one of a number of reasons but it is likely that they are less abundant than fin whales. Sei whales are not normally present in the Bay during the winter but return by May or June with their calves.

Fin Whale:

Distribution: Area 3.   Abundance: Considered to be a vulnerable species. Relatively rare worldwide.

Group Size: Generally 1-5 when travelling but may be seen in groups of up to 100 when on lucrative feeding grounds. Large groups in the area the ship passes through consist of 25- 50 animals.

Birth size: 6.4m(21.1ft)   Adult size: 17.5-26m(57.7-85.5ft)

The second largest animal living on the earth. Fin whales are often seen in the evening because the light in the evening helps to show the whales characteristic blow, which is tall (4-6m) and straight, (depending on the wind speed and direction). Fin whales were hunted commercially and their population suffered great losses as a result. When the ship passes through a feeding group, high spouts of water can be seen from some distance and they can be all around the ship. Fin whales may be encountered within a few hundred metres of the ship and closer!

Fin whales are the most frequently recorded large whales in the Bay of Biscay and often seen in very large extended groups. On one survey the ship travelling at 20 knots travelled for 40 minutes with fin whales or their blows visible all around the ship. A quick calculation told us that this group occupied an area of at least 140 square miles.

Blue whale:

Distribution: Area 3.   Abundance: Very rare in European waters. Occasionally recorded in the deep water of the Bay of Biscay. The world-wide population is thought to be between 6-14,000 individuals.

Group size: Generally 1-2 but may gather in larger numbers at feeding grounds. During 1996 a very large group was seen to gather off the California coast. One estimate put the numbers in thousands. Only a few hundred are thought to be left in the north Atlantic.

Birth size: 6-7m(19.8-23.1ft)   Adult size: 20-27m(66-88.5ft)

The blue whale is the largest animal on earth, it was devastated by whaling. When diving it usually raises its tail out of the water.

The blue whale usually dives to depths of around 150m but can go much deeper. Its sheer size is the most obvious identification feature.

Once an adult blue whale is seen it is an image stuck fast in the mind of the observer. It appears that a small island is rising from the ocean before your eyes and then disappearing again. The blow produced by a blue whale is very tall, straight and has an awesome power to it. These animals dive deep and use very low frequency sounds to communicate over vast distances some suggest around the entire world.

Use of photographs as part of the survey effort:

Some of these animals are so difficult to identify at sea that even photographs have to be scrutinised for many hours and even then a positive identification cannot be made. There are a number of species of beaked whale that may occur in the area but these are so elusive and little understood that photographs may be the only way to identify them. Even then the photographs will have to be viewed by many experts around the world to be sure of a species identification.

Some other species that may be present but have so far not been identified in the study area are; True's and Blainville's beaked whales, pygmy sperm whale and melon-headed whale.

Chapter 6

Birds of the sea crossing

Portsmouth Harbour

From the bridge, panoramic views can be obtained over the mudflats and saltmarsh of Portsmouth harbour. This large natural harbour is extremely rich in wildlife and has been designated a Site of Special Scientific Interest (SSSI). The area has large numbers of gulls throughout the year, and in winter thousands of waders and wildfowl. Of importance several thousand brent geese may be visible in areas of saltmarsh. Brent geese are long haul visitors from Spitzbergen, which lies within the Arctic Circle.

Portsmouth Harbour is the only place common gulls are seen in any numbers on the voyage. They are easily picked out from black headed gulls by their slightly larger size and thick black wing tips, which are edged white. Of special interest look out for little egrets, which are medium-sized heron-like birds, that have an elegant snow-white plumage. More associated with Mediterranean climes little egrets are recent colonists to Britain, perhaps as a result of global warming.

The English Channel

The main species seen in the central part of the English Channel on returning from Spain, are common sea birds such as gannet, kittiwake and large gulls, which are likely to have been encountered throughout the trip. The inshore waters of the Channel along the south coast of England provide more interest, especially in winter and wildfowl such as black and red throated diver, and black common scoter ducks may be seen. The English

Deciding how many animals you have seen can be very tricky as one surfaces so another disappears.

As the sun disappears it is enchanting to see dolphins gently break the surface.

A fast travelling dolphin pierces the water after a leap and creates a water funnel behind it.

Rolf Williams will often spend long periods perched precariously on the bow, in order to obtain some of these spectacular photographs.

A mother with her young approach the ship.

Bottlenose dolphins are much larger than common dolphins and are the animals associated with dolphinaria. The scars are probably from interactions with other bottlenose dolphins.

Common dolphins are the species most frequently seen in the bay and they have the hourglass pattern of white yellow and black.

The distinctive white markings on the head of the older Cuvier's beaked whale may help us to identify the whale.

Common dolphins bow-riding just in front of the ship's bulbous bow.

Dolphins look small and vulnerable but they are in control even in front of the largest ships.

A great shearwtaer - these birds travel around the entire Atlantic rim during the annual migration

A Cory's shearwater skims the wave tops.

Kittiwakes have a very attractive and delicate look.

An adult gannet gliding effortlessly.

A leaping Risso's dolphin shows itself by the very long pectoral fins, the tall dorsal fin, blunt head and 'peppered' colouration.

A pilot whale seen from the ship can sometimes be difficult to spot because they often move quite sedately and rarely breach the water surface.

We were surprised to see this merlin on the bow of the ship, but I suppose it is easier to take the ferry than fly!

Striped dolphins exploding from the sea like missiles.

Here in the wave a group of Risso's dolphins ride the surf.

Large greylag geese on migration, flying low over the sea.

Tuna leaping in the distance can often look like dolphins and it took some time before we could tell the difference.

Often the first and only indication that a whale is around is the blow or spout.

Some whale blows are very distinctive, like this sperm whale with its low bushy forward projected blow.

Here a juvenile gannet came to eye us up on the bridge.

Watching an 18 metre fin whale breaching even in the distance is an awesome sight.

This whale is around 20 metres long sometimes it can seem to be a very long time between the blow and the appearance of the dorsal fin.

There are three killer whales here, the last one has only got half a dorsal fin and this deformity may help us relocate it in the future years. (One child already has identified it as 'Willy" and was delighted to know that he had survived after his release).

Sunsets hold a particular fascination and magic at sea. There were no filters used to take this picture.

For each sighting an exact position, time, speed, and bearing for the vessel must be taken.

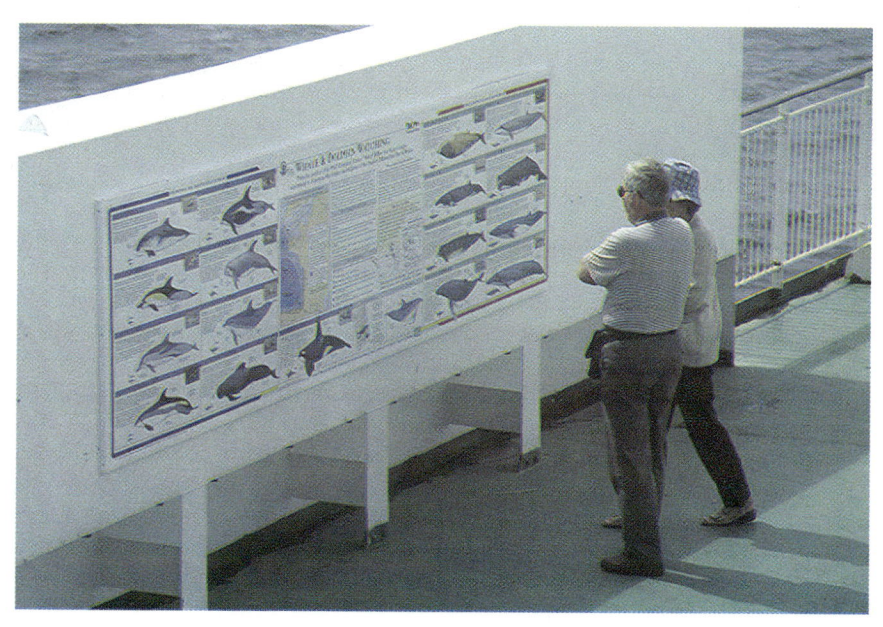

Passengers seeking to identify their sighting using the display board.

The observation deck quickly fills with passengers as the news of nearby dolphins spreads through the ship.

P&O European Ferries (Portsmouth) 'Pride of Bilbao'

Channel is also a good place to observe auks such as guillemot and razorbill.

Auks are curious looking birds, which nest on sea cliff ledges, where they faintly resemble penguins. Out at sea; look for them diving for cover immediately in front of the ship, as they are rudely awakened by the ship's passage.

The Continental Shelf around Brittany

Scanning out to sea in the shallow waters around the Brittany coast on the first morning of the southbound trip give the visitor early opportunities to familiarise themselves with the typical seabirds encountered over the journey. The first bird most likely to be seen and perhaps the quintessential seabird of the trip is the gannet.

Gannets are majestic and spectacular seabirds with spear-shaped bills. Adults are mainly white with black wing tips and yellow heads, whilst immatures are patterned with varying combinations of brown and white. Gannets are among the world's largest seabirds with a wingspan of nearly 2 metres. Watch out for them undertaking spectacular plunge-dives headfirst into shoals of fish, from heights of more than 100 feet.

The other main species likely to be seen, in this area is the kittiwake, which is a small gull patterned grey, black and white. Kittiwakes are readily identified from a distance by their buoyant, bounding flight. A number of other gull species similarly have grey wings with black tips, but only the kittiwake has all-black 'dipped-in-ink' tips. As their name suggests a good place to look for them is at the back of the boat, where individuals may follow the wake for an hour or more.

Kittiwakes are not the only gulls to trail the ship's lonely course, and the wake entourage frequently includes various scavenging species such as the herring gull and the great black-backed gull. The latter species is particularly distinctive being readily identified by its huge size, and jet-black wings.

North Biscay - Where deep and shallow waters meet

The north part of the Bay of Biscay, viewable from mid-afternoon onwards on the first full southbound day at sea, is perhaps the most interesting area of all for seabirds. In the north bay, the shallow continental shelf drops gradually towards the deeper water, and associated with this is nutrient upwelling, which creates an abundant supply of marine food. North Biscay is a particularly important feeding ground for large numbers of seabirds between August and early October.

The best place to look for birds at this time of year is around boats fishing for anchovies, cod, tuna and sardines. It is not uncommon to observe several hundred seabirds swirling around the wake of a single fishing boat, especially when the nets are being hauled up. Specialities include little and sabines gulls, great, sooty and manx shearwaters, arctic skua, pomarine skua, grey phalarope leaches, Wilson's and storm petrels.

Little and sabines are amongst the most beautiful of all gulls. Little gulls are extremely dainty. Adults have all grey upper wings which are blackened below, whilst immatures resemble kittiwakes. Sabines gulls are also similar to kittiwakes, but have black bills tipped yellow and pure black heads. Their forked tails make them extremely graceful and buoyant in flight, like a tern.

Shearwaters resemble small slender gulls, and can be picked out by their rapid, gliding flight. Calm weather is needed to see petrels, which are small and dark. The most frequently seen is the storm petrel, which is black with a white rump and is not much bigger than a sparrow.

Skuas are menacing birds, which prey on or steal from other sea birds. The most commonly seen species is the fierce great skua or bonxie. This is an intimidating bird, which looks like a large immature gull, but has prominent white flashes in the outer wings and is more bulky with a huge barrel chest.

In April and May large congregations of northbound seabirds such as lesser black-backed gulls, kittiwakes and fulmars, may be encountered resting in large flocks on the sea in the north Bay of Biscay. If the ship passes through the middle of a large group of seabirds, it can make for a spectacular sight. Fulmars are grey and white seabirds, identified by their characteristic stiff-winged, gliding flight which gives them the appearance of miniature albatrosses. They are distinctive, but some care is needed to distinguish them from shearwaters.

Deep water of the Bay Biscay

Birds may be few and far between in the deep waters of the southern and central parts of the Bay of Biscay. On calm afternoons in the summer, during the return journey from Santurtzi it is usually possible to see more whales than seabirds, and several hours may go by with no more than a handful of birds sighted! However, the few birds that are encountered can often be amongst the most interesting and unusual.

This section of the bay is chiefly of interest as the most reliable area for a number of 'southern' seabirds uncommon in British waters. Specialities include small numbers of Mediterranean shearwater, Cory's Shearwater, great shearwater , Wilson's petrel and sabines Gull. Mediterranean shearwaters are perhaps the most likely to be seen of these species. They resemble British manx shearwaters, but have a uniformly dark brown plumage, rather than one that is black and white.

Mention must also be made of the Wilson's petrel, which is one of the most abundant birds in the world, but a rarity in British waters. It has been reported from this area on several occasions between July and September. Look for a slightly oversized storm petrel, which paddles its feet on the surface of the water when feeding.

Bird interest largely occurs in the warmer months, but on short winter days it occasionally possible to locate small numbers of puffins and other auks. puffins or 'the clown of the cliffs' are among the most familiar of all sea birds, because of their comical, over-sized gaudy bill. However, they can be quite difficult to identify when in drab winter plumage. They are best told from other auks such as the guillemot in flight, where they look much dumpier, and flap their wings more rapidly.

Land Birds at Sea

A feature of trips especially during the spring and autumn months, are the regular sightings of land birds on migration. It may come as a surprise that more land than seabirds have so far been seen out at sea. The long list of species seen far out at sea even includes birds of prey such as hobby, kestrel, merlin and sparrowhawk.

On one October trip we were lucky enough to observe the bizarre sight of a merlin catching, then eating a meadow pipit in the middle of the Bay of Biscay!

The majority of migrants observed are passerines (perching birds) such as finches, warblers, pipits, wagtails, hirundines, thrushes and chats. Many of these are seen to land and roost on the boat. The most varied months are April and October. Species which have been seen so far include familiar garden birds, such as starling, robin, blackbird and song thrush.

Less familiar species have included turtle dove, wheatear, garden warbler, black redstart and brambling. Turtle doves are particularly lazy and have been observed in spring hitching a lift for the full length of the northbound journey!

Land birds are generally seen in ones and twos, although hordes of swifts have been observed above the ship on a few occasions. In addition, members of the crew have occasionally noted very large arrivals ('falls') of small passerines on the boat, especially finches during inclement weather at migration times. At such times it appears that there are more birds on the boat than people.

Santurtzi Harbour

At a first glance Santurtzi harbour can seem an unlikely looking spot for the ornithologist, but on close inspection there are species of interest throughout the year.

The yellow-legged gull is a speciality of Santurtzi harbour. This species is very similar to the closely related herring gull, which is the characteristic noisy 'sea gull' found in every seaside locality in Britain. On close examination yellow-legged gull adults are most easily separated from herring gulls by their slightly darker mantle (back colour), and as their name implies by having yellow rather than pink legs.

Yellow-legged gulls are present throughout the year in the harbour, and close views can be obtained during the day in the waters around the ferry terminal. Small flocks are often observed following fishing boats in and out to sea, but the largest numbers (usually 1-2000) can be seen on breakwaters or on flat roofed building yards of the port especially early morning or late afternoon.

Small numbers of the familiar 'playing field' black headed gull usually occur in the harbour throughout the year and a careful search of their groups will often reveal one or two scarcer Mediterranean gulls. These beautiful gulls can be picked out from black headed gulls by their slightly larger size and pure pearly-white wings.

The other main bird species encountered in the harbour are cormorants and shags, with mixed groups of up to 70 being present during the winter months. The best place to look for them is on concrete marker buoys in the middle of harbour, where they may often be seen in discrete groups of up to 20. Look for birds hanging their wings out to dry in the first late morning sun. This is a most curious activity and follows periods of deep diving for fish. Cormorants are a truly coastal bird and they are rarely seen more than a few miles out to sea.

During the winter, the harbour provides shelter for a variety of seabirds in small numbers including auks such as the guillemot, ducks such as common scoters, grebes and divers. The spring and summer seasons provide much of interest for the bird watcher as a variety of migrant sea and coastal birds may pass through the harbour. Our trips so far have produced a diverse range of species including avocet, stone curlew, and black tern.

Chapter 7

Wildlife around Santurtzi

About a ten-minute walk from the ferry terminal through the narrow streets of Santurtzi, are a series of low hills.

These provide a stark and picturesque contrast to the grim industrial port settings. Owing to the steep terrain these foothills have escaped the ravages of modern 'industrial' farming practices, and the landscape is largely maintained by traditional forms of grazing and cultivation.

An interesting variety of habitants occur over a relatively small area and includes small terraced fields, allotments, scrub, acacia and sweet chestnut groves, eucalyptus plantation, hay meadows and rough pasture grazed by sheep, cattle and goats. In addition, there is a small amount of pine forest and open areas with extensive bracken and gorse.

This exceptional diversity of habitats supports a wealth of species, including many which are declining throughout Europe as farming practices continue to change at the expense of wildlife.

A special bird of this archaic landscape is the red backed shrike, which is common during the spring and summer months along unkempt, scrubby field edges. Males of this species are extremely handsome with their black mask, rich chestnut back and thick hooked falcon-like beak. Although little bigger than a sparrow, shrikes or *butcherbirds* as they are colloquially known, are birds of prey. They have the curious habit of impaling their quarry such as other small birds on a thorn bush prior to devouring them.

Scrub, rough grassland and open woodland areas are important for a variety of other 'southern' species which are uncommon or rare in Britain including melodious warbler, wryneck, firecrest, Sardinian warbler, fan-tailed warbler, nightingale and cettis Warbler.

A great many more species pass through and spring bird watching can be particularly exciting as bushes and field may team with northward-bound grounded migrants such as hoopoe, ortolan bunting, tree pipit, pied flycatcher, icterine warbler and orphean warbler.

At such times look out for red-rumped swallow and alpine swift amongst flocks of more familiar hirundines such as swallow, house martin and swift.

Throughout the area it is worth looking up into the sky at regular intervals as birds of prey are regularly seen overhead at all times of the year. Species present include red kite (winter), kestrel, goshawk, buzzard and even griffon vulture.

Most of the small fields in the area have escaped applications of fertilisers and herbicides and consequently are ablaze with a dazzling array of pretty wildflowers. There is interest year-round - spring comes early to Santurtzi, but in summer the cloudy mornings and high humidity ensures a prolonged and colourful flowering season.

In rocky base-rich areas where grazing intensity is high and the grass short, plant species diversity is particularly rich, with species such as sombre bee orchid, tongue orchid, yellow bee orchid, white rockrose and green hellebore occurring. Areas of scrub are equally rich and the impressive array of species includes hepatica, blue gromwell and many types of shrub. In early summer, arable and allotment fields are choked with weeds such as corncockle and field eryngyo.

In such a colourful and flower-rich landscape, butterflies are equally abundant including a number of species not found in Britain; such as Cleopatra, large chequered skipper, ilex hairstreak, long-tailed blue and scarce swallowtail. Of these, the large chequered skipper is a local speciality, as it is rare in Spain being restricted to a narrow part of the coastal strip between Ovieda and San Sebastian.

Search for the butterfly in June and July around partially shaded grassy roadsides. Of further interest the large blue, which is a species of European Conservation importance, also occurs in small numbers.

Many other species of insect occur, including of special interest, the praying mantis, hummingbird hawkmoth and the bizarre-looking mole cricket.

The area has a high density of seed eating birds. Late summer sees large noisy post-breeding of mixed flocks of finches and other small birds flitting between meadows and other patches of rough ground – a sight once familiar across Britain 20 years ago. Amongst the large flocks of linnet, greenfinch, goldfinch and house sparrow, look closely for less common species such as the tree sparrow and serin.

This is heaven for the pure naturalist.

Chapter 8

Marine Conservation and the Biscay Dolphin Research Programme

It would be quite reasonable to ask 'why do three people travel every month for four days to Spain giving up their annual leave and time off'. It is difficult to explain the need to carry out a long- term project of this kind. Conservation means different things to different people.

To me conservation means to preserve and protect. All too often humankind destroys to the point of extinction and then tries to pull back from that ultimate destruction. This is not conservation.

It was only brought home to me when I experienced at first hand the rich and varied cetacean life of the Bay of Biscay. Here was a wondrous environment, accessible and yet at the same time undiscovered.

When we started, our aim was purely scientific; to find out what cetaceans are in the Bay and their relative abundance. As time progressed and we encountered large fishing fleets, naval vessels on manoeuvre and suprisingly large amounts of rubbish we began as one to think about what we were experiencing.

We compiled our records and compared them against descriptions of whale watching holiday encounters and other research work around the world. It became clear that this was an area of almost unrivalled splendour. As we told others of our experiences we heard the same thing time and time again, 'I didn't know there were whales and dolphins around here'.

Many years ago I was guilty of the same error when here, all the time, the truth was so different.

The Bay of Biscay is an area of outstanding beauty and character; it acts as home and temporary shelter to a wide range of birds, cetaceans and other life. It is a maternity ward and nursery, it is a restaurant and playground and it is one of the world's greatest 'motorways'.

Conservation is to preserve this diversity, to protect this variety and to respect the environment where all this can be found.

Through our work we will provide the baseline information against which changes can be monitored. I hope that we will raise awareness of this area and the life within it. Conservation is education, if people are given the opportunity to see for themselves free ranging ocean going birds and cetaceans then the sheer scale and wonder of their lives will touch their hearts and they will become the protectors.

Humanity has here a unique opportunity to observe a rich and diverse environment and to keep it that way. We hope that our work will help to conserve the Bay of Biscay and its inhabitants.

Cetaceans recorded February 1998

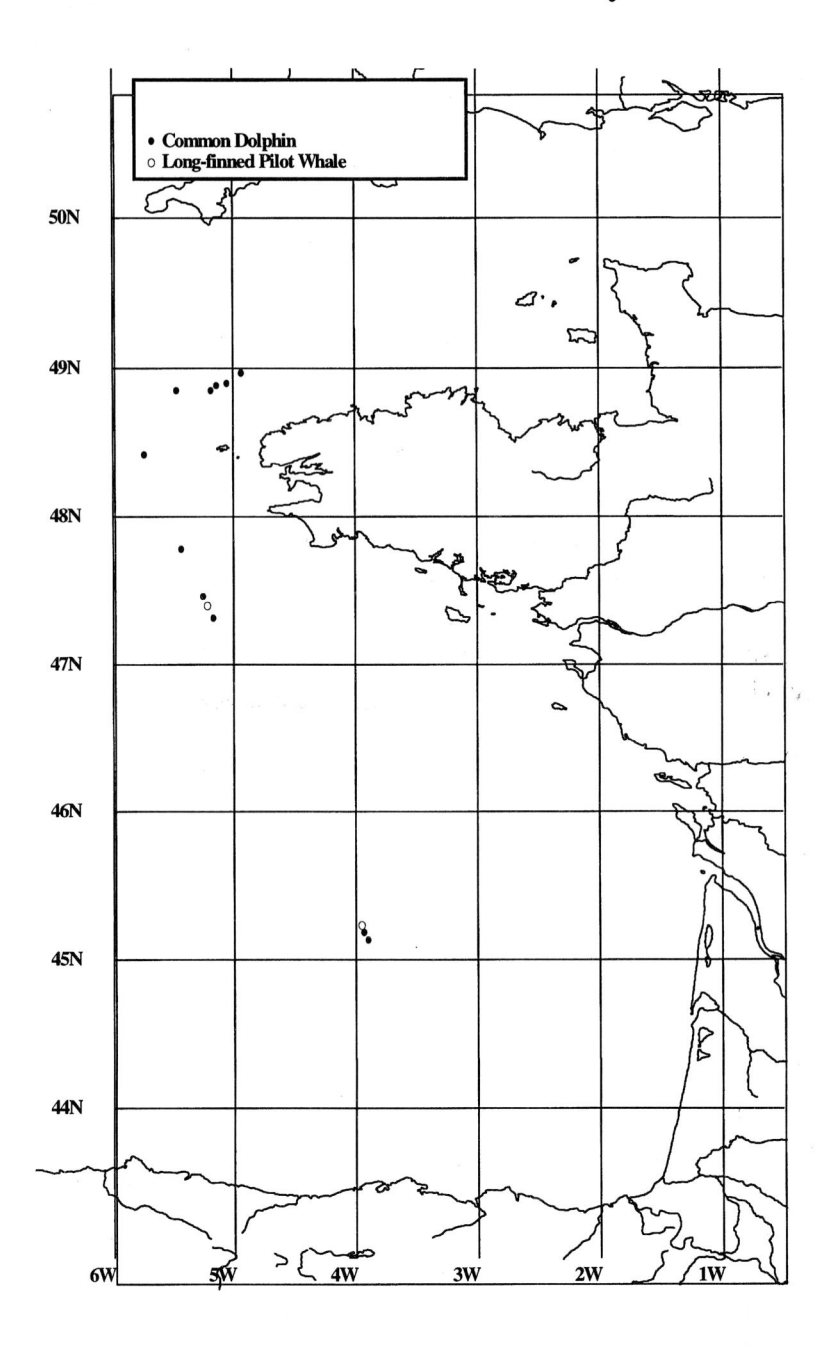